STUDENT GUIDE

BUYER BEWARE

RATES, RATIOS, PROPORTIONS, AND PERCENTS

MathScape™
SEEING AND THINKING MATHEMATICALLY

To: Staff Reporters
From: Buyer Beware Magazine, Inc.

Welcome to Buyer Beware.
As a staff reporter, your job is to help your readers become more educated consumers. Mathematics is an important part of consumer awareness. You will use unit pricing to find the better buy, write ratios to compare brands, and use proportions to increase recipes and determine cost. You will use percents to estimate discounts and interpret data in circle graphs.

What math is involved in being an educated consumer?

BUYER BEWARE

PHASE**ONE**

Rates

In this phase, you will compare the prices and sizes of products to determine which is the better buy. Next you will use a price graph to find and compare unit prices. Then you will decide when it is cheaper to buy by the pound. Finally, you will test claims made by a sandwich bar to see whether it is less expensive to buy a sandwich there or to make one at home.

PHASE**TWO**

Ratio and Proportion

You will begin this phase by writing ratios to compare quantities. Then you will use a ratio table to find equal ratios. Next, you will use equal ratios to compare brands. Then you will use proportions to solve problems in a variety of situations and determine what makes a situation proportional. You will end the phase by using what you know about ratio and proportion to create a mosaic design.

PHASE**THREE**

Percents

You will begin this final phase by using familiar benchmarks, then counting and rounding to estimate expenses in a budget. Next, you will interpret and create circle graphs representing budgets for a drama club. Then you will use percents and discount coupons to find your savings in a percent-off sale. You will end the unit through a final activity: planning an athletic banquet on a budget.

PHASE ONE

To: Staff Reporters
From: The Editors

Your first assignment is to test claims made about unit price. Super Sandwich Bar claims that it is less expensive to make a sandwich at their sandwich bar than it is to make one at home. When you complete the assignment we would like you to write an article for *Buyer Beware* stating your findings.

To prepare for this assignment, you will need some experience working with rates and unit prices.

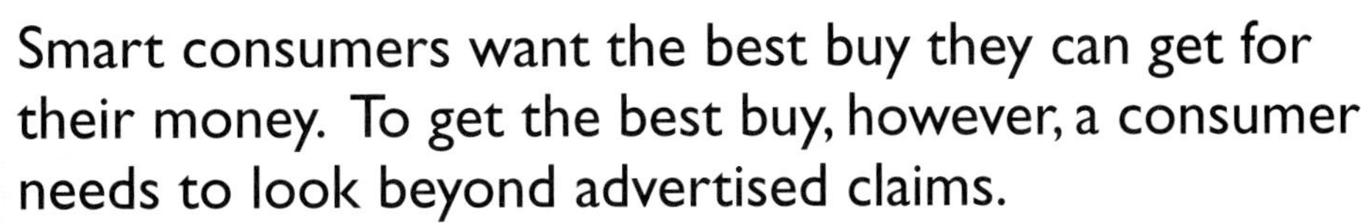

Smart consumers want the best buy they can get for their money. To get the best buy, however, a consumer needs to look beyond advertised claims.

In order to do accurate comparison shopping, you need to know how to calculate unit price. By finding the unit price—the price per ounce, pound, or item—you will learn to make informed decisions and find the products that are better buys.

Rates

WHAT'S THE MATH?

Investigations in this section focus on:

DATA and STATISTICS

- Finding unit prices using a price graph
- Constructing a price graph to compare unit prices

NUMBER

- Comparing unit prices of different-size packages
- Comparing unit prices of different brands
- Finding the price per pound to decide the better buy
- Comparing prices per pound
- Calculating long-term savings
- Calculating unit prices and total prices

mathscape2.com/self_check_quiz

What's the Best Buy?

COMPARING UNIT PRICES

Shoppers need to be able to calculate unit prices to find the best buy. In this lesson, you will compare various-size packages of cookies made by the same company to decide which size is the best buy. Then, you will compare two different brands of chocolate chip cookies to decide which one gives you more cookie for your money.

Compare the Unit Prices of Different-Size Packages

How can you compare cookie packages to find the best buy?

The Buyer Beware Consumer Research Group has collected data on chocolate chip cookies. They want you to find out which package of Choco Chippies is the best buy.

To find the best buy, you need to find the unit price, or the price per cookie, for each size package. You can find the price of one cookie in a package if you know the total amount of cookies in the package and the price of the package.

1. Use a calculator to figure out the price per cookie for each package. Round your answers to the nearest cent.
2. Decide which package size is the best buy. Explain how you figured it out.
3. List the different Choco Chippies package sizes in order from best buy to worst buy.

Choco Chippies Prices

Package Size	Number of Cookies	Package Price
Snack	4	$0.50
Regular	17	$1.39
Family	46	$3.99
Giant	72	$5.29

Compare the Unit Prices of Two Different Brands

How can you use unit price to determine which of two brands is the better buy?

At *Buyer Beware* magazine we frequently get letters from our readers asking questions about best buys. Here is one of the letters we received:

> Dear Buyer Beware,
>
> Help! My friend and I don't agree on which brand of cookies is the best buy. She's convinced that it's Mini Chips, but I'm sure it's Duffy's Delights. Which is really the better buy?
>
> The Cookie Muncher

The research group at *Buyer Beware* has put together Cookie Prices data for you to use.

1 Decide which unit you would use to compare the two different brands of cookie. Explain why you chose that unit.

2 Find the price per unit of each brand of cookies.

3 Compare the unit price of the two brands of cookies. Is one brand the better buy? If so, explain why. If not, explain why the brands are equally good buys.

Cookie Prices

Brand	Package Price	Number of Cookies	Package Weight
Mini Chips	$1.39	17	6 oz
Duffy's Delights	$2.29	10	11 oz

Determine the Better Buy

Write a response letter to The Cookie Muncher.

- Describe what you did to figure out the better buy.
- Give evidence to support your conclusions.
- Give some general tips for finding the best buy.

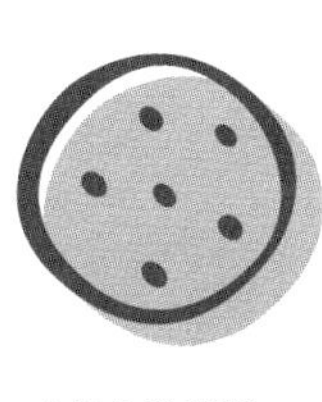

MINI CHIPS

DUFFY'S DELIGHT

hot words
unit price
rate
Homework
page 34

2 The Best Snack Bar Bargain

INTERPRETING AND CREATING PRICE GRAPHS

You can use a price graph to compare unit prices for different products. In this lesson you will use a price graph to determine the price at different quantities of a snack bar if you were paying by the ounce. Then you will construct a price graph to compare the prices of five different products.

Use a Price Graph to Find Unit Price

How can you use a price graph to estimate snack bar prices?

The graph below shows the prices for three different snack bars. The price of Mercury Bars is $1.00 for 2 oz. Jupiter Bars are $2.98 for 3.5 oz and Saturn Bars are $3.50 for 4.5 oz.

Each of the three dots on the graph shows the price and the number of ounces for one of the snack bars. Each line shows the price of different quantities of the snack bar at the same price per ounce.

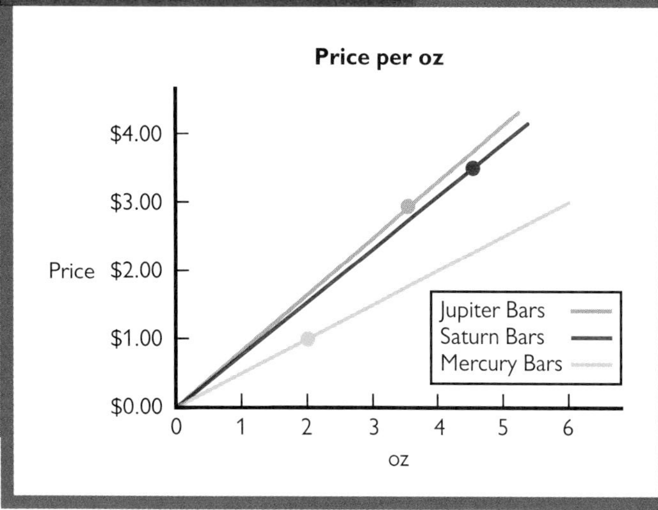

1. Use the price graph to find the price of a 3-oz Mercury Bar.
2. Use the price graph to find the price of a 0.5-oz Saturn Bar.
3. Use the price graph to find which snack bar has the lowest unit price and which has the highest unit price.
4. Use your calculator to find the unit price of 1 oz of each of the snack bars. Use the price graph to check your calculations.

Construct a Price Graph to Compare Unit Prices

How can you use a price graph to compare product prices?

The Buyer Beware research group has collected data on five different products.

Product	Number of Units	Price
Oatmeal	14 oz	$2.40
Tuna	7.5 oz	$2.00
Penne pasta	8 oz	$1.00
Sourdough pretzels	12 oz	$2.60
Whole wheat rolls	9 oz	$1.60

1 Use the above data to construct a price graph.

2 Use the price graph to complete a Price Comparison. For each product, record the price for 1 oz, 3 oz, 4.5 oz, and 6 oz.

Price Comparison Table

Product	1 oz	3 oz	4.5 oz	6 oz
Oatmeal				

3 Which product is the most expensive per ounce? the least expensive per ounce?

Write About Your Price Graph

Think about what you have learned about interpreting and making price graphs.

- Write a description of your price graph and what it shows. Explain how to use your price graph to find prices for packages that are larger and smaller than your original package.
- Describe how two products will look on a price graph if their prices are almost the same per ounce.

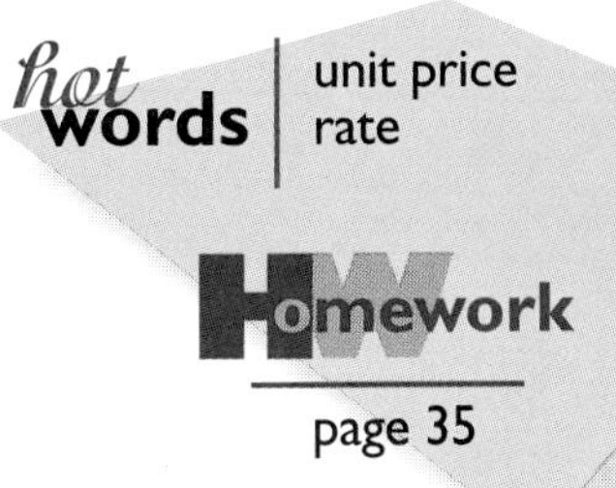

hot words
unit price
rate
Homework
page 35

3 Cheaper by the Pound

USING UNIT PRICES AND WEIGHT UNITS TO COMPARE PRICES

Sometimes buying in "bulk" or larger quantities will save you money. In this lesson you will find the price per pound of different-size packages of rice to decide which one is the best buy. Then you will evaluate the price per pound of items at a silly sale.

How can you find the package that is the least expensive per pound?

Find the Price per Pound to Decide the Best Buy

1. What is the price per pound of each package of rice? Round your answer to the nearest cent.
2. Find out which package of rice is the best buy in terms of price per pound.
3. Explain which package of rice makes the most sense to buy if only one person in a family eats rice.
4. Decide which package of rice would be the best buy for your family.

Fluffy Rice	
2 lbs	$1.09
5 lbs	$2.69
10 lbs	$4.99
20 lbs	$7.99

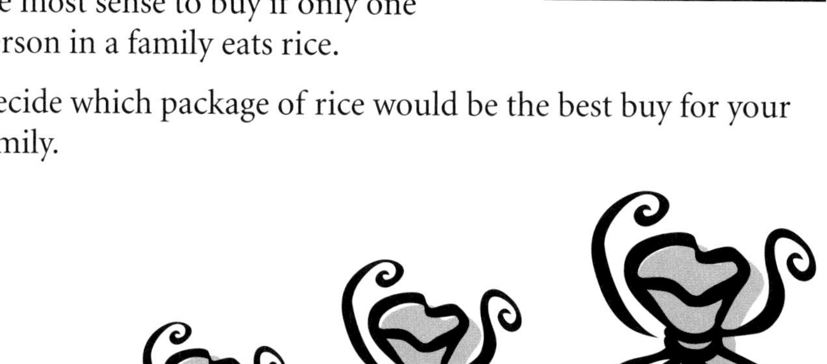

2 POUNDS
$1.09

5 POUNDS
$2.69

10 POUNDS
$4.99

20 POUNDS
$7.99

Compare Prices per Pound

Suppose you go to a silly sale where everything is rated by price per pound. Which is cheapest per pound: a bicycle, a pair of sneakers, a video camera, or a refrigerator?

How can you determine the prices per pound of different items?

43 POUNDS
$139.99

2 POUNDS
$84.99

11 POUNDS
$799.00

230 POUNDS
$679.99

1. Use your calculator to find the price per pound of each item.
2. Rank the items from the least to most expensive per pound.
3. What items have high prices per pound?
4. What items have low prices per pound?

Write About Buying in Bulk

Write what you know about buying in bulk. Be sure to answer these questions in your writing:

- How can you figure out if the largest size is the least expensive per pound?
- Is the largest size always the least expensive per pound?
- When is it a good idea to buy products in bulk? When is it not a good idea?
- Could the same purchase be a good choice for one consumer but not for another consumer? Explain.

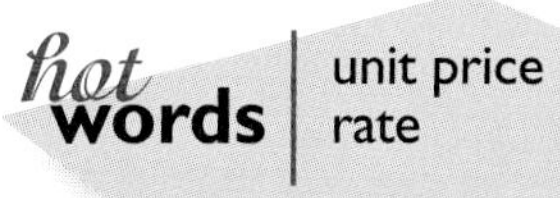

page 36

4 It Really Adds Up

SOLVING REAL-LIFE PROBLEMS AND FINDING THE BETTER BUY

In this lesson you will use what you have learned about unit prices to solve real-life problems. First, you will find and compare the prices of two different brands of pretzels. Then you will take on the role of an investigative reporter to test the claim made by a sandwich bar.

Calculate Long-Term Savings

How much can you save over time by buying a less expensive brand?

People are often surprised at how buying a little snack every day can really add up over time. The Buyer Beware research team wants you to figure out the price of buying two brands of pretzels for different time periods.

1. Find the price of buying one bag of pretzels every day for a week, a month, and a year. Make a table like the one below to organize your answers.
2. How much would you save if you bought No-Ad Pretzels instead of Crunchy Pretzels for the different time periods?
3. If you bought a bag of No-Ad Pretzels every day instead of Crunchy Pretzels, how many days would it take to save $30? $75? Explain how you figured it out.

Pretzel Costs

Type of Pretzels	Price for 1 Bag a Day for 1 Day	Price for 1 Bag a Day for 1 Week	Price for 1 Bag a Day for 1 Month (4.3 weeks)	Price for 1 Bag a Day for 1 Year
Crunchy Pretzels	$0.65			
No-Ad Pretzels	$0.50			
Savings for buying No-Ad Pretzels				

Calculate Unit Prices and Total Prices

At Super Sandwich Bar, customers can make their own Terrific Ten Sandwich—two slices of cheese and eight slices of meat—for only $3.59. The restaurant claims that this is cheaper than making the sandwich at home. The Buyer Beware research group wants you to test this claim. They have collected information on the prices of different ingredients for you to use.

How can you test a claim by figuring out unit prices?

Sandwich Tips

STEPS TO DESIGNING A SUPER SANDWICH

1. Choose *at least* three ingredients from the handout In Search of the Terrific Ten Sandwich.
2. Make a table. List the ingredients you chose, the amount of each, and the price of each.
3. What is the total price of your giant sandwich if you paid for it by the slice?
4. Do you think Super Sandwich Bar's price of $3.59 is a good deal? Explain your reasoning.
5. Name your sandwich and draw a picture of it for a magazine article.

Write About Super Sandwich Bar's Claims

- Describe the strategies and solutions you used to figure out the total price of your sandwich.
- Write a magazine article discussing your findings about the claims made by Super Sandwich Bar. In your article answer the question: Is it cheaper to make a sandwich at the bar or buy all the ingredients and make it yourself?

PHASE TWO

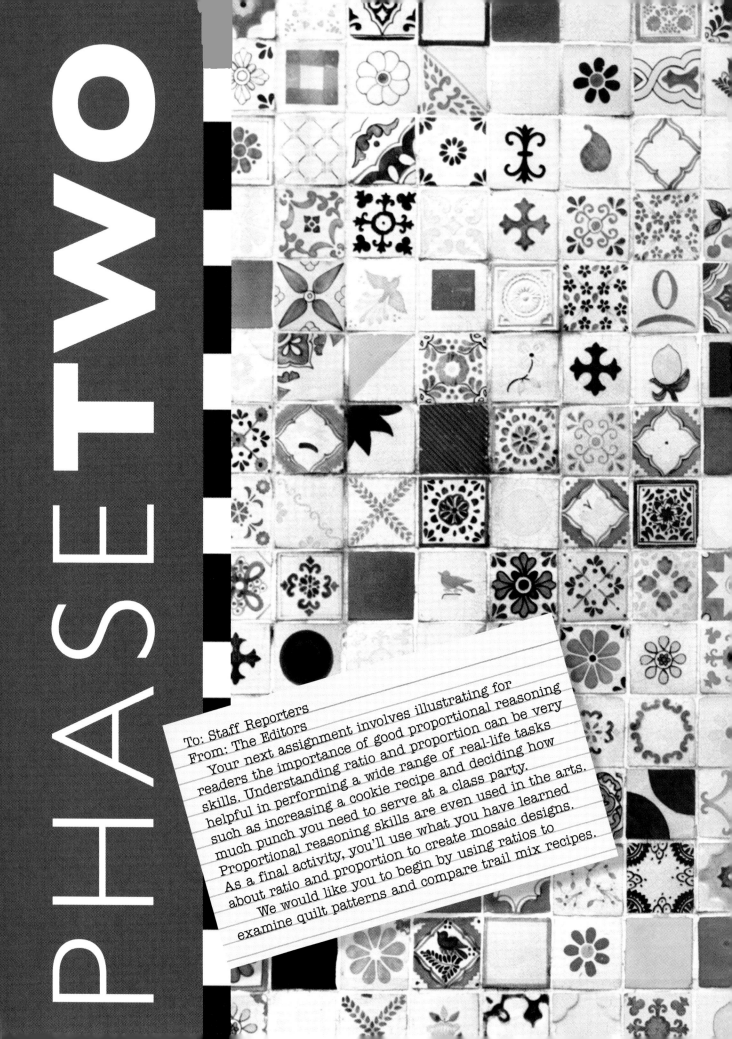

To: Staff Reporters
From: The Editors

Your next assignment involves illustrating for readers the importance of good proportional reasoning skills. Understanding ratio and proportion can be very helpful in performing a wide range of real-life tasks such as increasing a cookie recipe and deciding how much punch you need to serve at a class party. Proportional reasoning skills are even used in the arts. As a final activity, you'll use what you have learned about ratio and proportion to create mosaic designs.

We would like you to begin by using ratios to examine quilt patterns and compare trail mix recipes.

Can ratio and proportion be useful in planning school activities? A ratio is a relationship between two quantities of the same measure. Ratios are useful when you want to compare data, decide quantities of servings, or compare different products.

A proportion states that two ratios are equal. Setting up and solving a proportion can help you increase recipes or create a design to certain specifications.

Ratio and Proportion

WHAT'S THE MATH?

Investigations in this section focus on:

NUMBER

- Using ratios to compare data
- Using equivalent fractions to compare ratios

SCALE and PROPORTION

- Finding equal ratios with a ratio table
- Using cross products to compare ratios
- Using equal ratios and cross products to solve proportions
- Writing and using proportions to solve problems
- Determining when to use proportions to solve a problem

5 Quilting Ratios

WRITING AND COMPARING RATIOS

A ratio is a comparison of one number to another by division. In this lesson, you will use ratios to describe a design. Then you will use ratios to compare trail mix recipes.

How can you use part-to-part and part-to-whole ratios to describe a design?

Practice Writing Ratios

Leticia is designing a quilt pattern that is made up of squares. One row of the quilt is shown below.

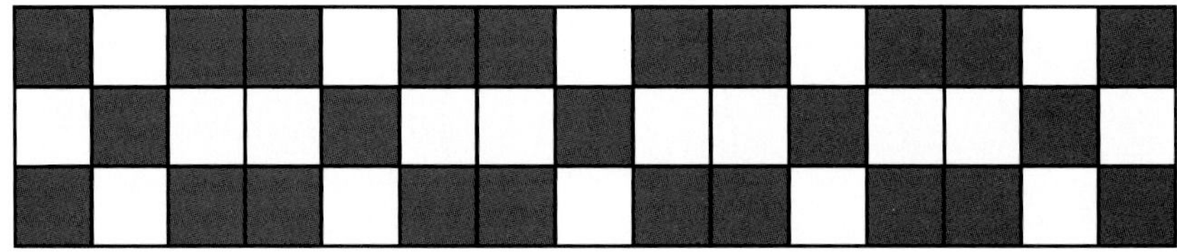

1 What is the ratio of blue squares to yellow squares? Is this a part-to-part or part-to-whole ratio?

2 What is the ratio of blue squares to all the squares in the row? Is this a part-to-part or part-to-whole ratio?

3 Gino is designing the quilted place mat at the right. Find two different ratios within the design. Describe them using words and numbers. Tell whether they are part-to-part or part-to-whole ratios.

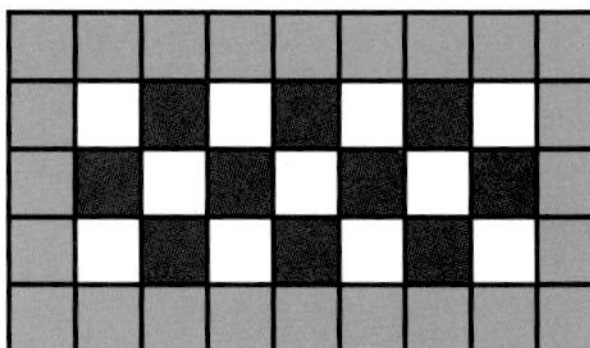

Writing Ratios

You can write ratios in several ways. For example, in one classroom, there are 2 boys for every 3 girls. This ratio can be written as follows.

2 to 3 2:3 $\frac{2}{3}$

These are part-to-part ratios. *There are 2 boys for every 3 girls in class.*

This class could also be described using a part-to-whole ratio. *There are 2 boys for every 5 students in class.*

Compare Ratios in Trail Mix

How can comparing fractions help you decide which trail mix to buy?

Dwight Middle School is planning an Autumn Festival. The seventh grade is planning to buy trail mix to sell at the festival. They want to decide if one of two brands, Mountain Trail Mix or Hiker's Trail Mix, contains more chocolate by weight or if both contain the same amount. The *Buyer Beware* research group has collected the information shown on the two brands. Note: The mixes contain only dried fruit and chocolate chips.

1 What is the ratio of chocolate chips to dried fruit by weight in Mountain Trail Mix?

2 What is the ratio of chocolate chips to dried fruit by weight in Hiker's Trail Mix?

3 Are the ratios of chocolate chips to dried fruit for each type of mix the same or different?

4 Does one trail mix brand contain more chocolate per ounce than the other? If so, which one? How do you know?

Mountain Trail Mix
6 oz chocolate chips
10 oz dried fruit
Hiker's Trail Mix
4 oz chocolate chips
6 oz dried fruit

Write Ratio Statements

Write five different ratio statements about Mountain Trail Mix. Here are two examples to help you get started.

> The ratio of chocolate chips to dried fruit is 6:10.
>
> There are 3 ounces of chocolate chips for every 5 ounces of dried fruit.

- For each ratio statement, label it either part-to-part or part-to-whole.
- Add one additional ingredient to the Mountain Trail Mix so that you can create either a part-to-part or a part-to-whole ratio of 2 to 3. How many ounces of the ingredient will you add? Possible ingredients include peanuts, coconut, pretzels, or cereal.

hot words
ratio
equivalent fractions
HW Homework
page 38

6 In the Mix

USING RATIO TABLES

Ratio tables can help you organize information, identify a pattern and extend it. In this lesson, you will use ratio tables and other methods to create batches of fruit punch that maintain a consistent fruitiness.

How can you use ratios to increase recipes?

Explore Methods for Increasing Quantities

There are many kinds of drinks you can buy at the store in the form of a liquid concentrate to which you add water. For example, one kind of fruit punch uses 3 cans of water for each can of fruit punch concentrate.

1 How many total cans of liquid would you use to make 2 batches of fruit punch? Each batch uses 1 can of fruit punch concentrate.

2 How would you create a bigger batch of punch that uses a total of 20 cans of liquid (both concentrate and water)? The punch should have the same amount of fruity taste as a single batch does. How many cans of fruit punch concentrate would you use? How many cans of water would you use?

Students are making punch for a class party and need to serve everyone who will be coming. From last year's party, the students estimate they will need to make 3 batches of punch.

3 How many cans of fruit punch concentrate will they need?

4 How many cans of water will they need?

Be ready to describe your thinking.

Compare Solutions

Three students explained how they created three batches of punch for the class party. Do you agree or disagree with each student's reasoning? If you disagree, explain why.

How can ratios help you solve a problem?

Kira

In one batch of the punch, 3 of the 4 cans are water, so $\frac{3}{4}$ of the punch is water. In my bigger batch, I'll need $\frac{3}{4}$ of the total to be water. If I triple the recipe, there will be a total of 3×4 or 12 cans of liquid in all. Since $\frac{3}{4}$ of 12 is 9, I'll use 9 cans of water and 3 cans of concentrate.

Here's Kira's picture.

Here's the bigger batch.

Tobi

The directions call for 1 can of concentrate and 3 cans of water. Since $3 - 1 = 2$, there are 2 more cans of water than cans of concentrate. In my bigger batch of punch, I'll need twelve cans in all. So I'll use 5 cans of concentrate and 7 cans of water.

Here's Tobi's picture.

Mitchell

I'll use a pattern to figure this out. I know I need 3 cans of water for each 1 can of concentrate. I'll just keep adding 1 can of concentrate and 3 cans of water until we have three batches.

Here's Mitchell's picture.

Halftime Refreshments

UNDERSTANDING PROPORTIONS

A proportion shows that two ratios are equal. In this lesson, you will use proportions to decide how many spoonfuls of hot cocoa mix are needed to make mugs of cocoa. Then you will use proportions to solve problems in other contexts.

Use Different Methods to Solve Proportions

How are proportions related to ratios?

Proportion

A proportion is a comparison between two equal ratios. Often, a proportion is written as two equivalent fractions. For example, $\frac{12 \text{ inches}}{1 \text{ foot}} = \frac{36 \text{ inches}}{3 \text{ feet}}$.

What other proportions can you think of?

Set up a proportion and solve each problem below.

1. The seventh grade is planning to sell mugs full of hot cocoa at the football game. If 6 spoonfuls of cocoa mix make 3 mugs of hot cocoa, how many spoonfuls are needed to make 9 mugs?

2. How many spoonfuls would be needed for:

   a. 21 mugs? **b.** 36 mugs? **c.** 96 mugs?

3. How many mugs would you get from:

   a. 14 spoonfuls? **b.** 30 spoonfuls? **c.** 64 spoonfuls?

4. The seventh grade is going on a field trip to the local science museum. The school policy on field trips states that there must be one adult chaperone for every 8 students. How many chaperones are needed for:

   a. 40 students? **b.** 96 students? **c.** 120 students?

More Proportion Problems

How can you use what you know to solve problems using a proportion?

Solve each problem.

1. Owen is converting a cookie recipe to feed a large group of people. One of the ingredients his recipe calls for is 2 cups of flour. His recipe makes 3 dozen cookies. If he has 12 cups of flour, how many dozen cookies can he make?

2. Alecia is traveling in Canada with her family and notices that the road signs have distances in both miles and kilometers. At one point, the sign says the distance to Quebec is 65 miles or 105 kilometers. If their trip covers 250 kilometers in all, what distance will they have traveled, rounded to the nearest whole, in miles?

3. Jeanine is ordering pizzas to feed students who participate in the school car wash fund-raiser. At another school fund-raiser last month, 10 students ate 4 pizzas. How many pizzas should she plan to order if there will be 75 students participating?

4. Nurses sometimes use proportions when they take your pulse. A healthy heart rate is about 72 beats per minute. Some nurses take your pulse for 15 seconds, then estimate your heart rate. How many beats would they expect to count in 15 seconds if you have the average healthy heart rate?

Write About Proportions

Write your own problem that can be solved using proportions. Use the problems above as a model. Write the solution to the problem on a separate sheet of paper.

hot words
proportion
cross product
Homework
page 40

8 Can I Use a Proportion?

USING PROPORTIONS TO SOLVE PROBLEMS

In the lesson, you will examine real-life situations to determine whether they are proportional. Then, you will design a mosaic to show what you have learned about ratio and proportion.

How do you know when you can use a proportion to solve a problem?

Determine if a Situation is Proportional

Nellie's favorite trail mix has 4 ounces of peanuts and 6 ounces of chocolate. She wants to make a big batch with the same ratio of peanuts to chocolate to bring on a hike with her friends this weekend.

1 Complete the table to help Nellie make different sized batches.

Peanuts (ounces)	4	6	8	10	
Chocolate (ounces)	6				

2 On graph paper, plot each ordered pair from the table and draw a line through the points.

3 Is the ratio of peanuts to chocolate the same regardless of the size of the batch? Is this situation proportional? Explain.

Susan is going to a school fair. She has to pay $2 to enter the fair and $1 per ticket for games and refreshments.

4 Complete the table to help Susan determine her total expenses based on how many tickets she purchases.

Number of Tickets	4	6	8	10	
Cost (dollars)	6				

5 On graph paper, plot each ordered pair from the table and draw a line through the points.

6 Is the ratio of tickets purchased to total cost of the fair the same regardless of the number of tickets purchased? Is this situation proportional? Explain.

Write About Solving Proportions

How can you use what you have learned to solve problems?

A student designed the winning mosaic for the school's new entryway. Now the building committee needs to calculate the cost of her design given the following prices for color tiles.

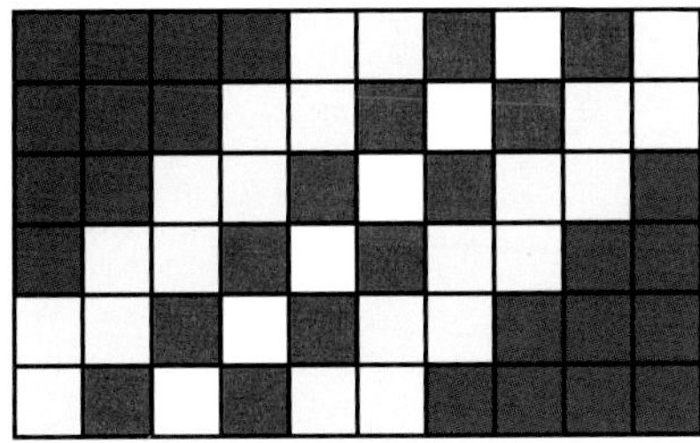

Blue tiles	5 for $4.00
Red tiles	5 for $6.00
Yellow tiles	5 for $3.00
White tiles	5 for $2.50

Refer to the handout Different Ways to Solve Proportions to help you answer each question.

1. Calculate the cost of blue, red, and yellow tiles in the mosaic design using a different method for each color. Write the name of the method you used and an explanation of how you calculated each cost using this method.

2. Calculate the total cost of the white tiles using any method you wish. Then write an explanation of how you calculated the cost using this method.

Design a Mosaic

In this activity, you will design a mosaic to show what you have learned about ratios and proportions.

Use the following guidelines to plan and create your design:

- Your design should have 3–6 different colors in it.
- Your design should have between 30–60 total squares of the same size in it.
- Two of your colors should be in a part-to-part ratio of 2:3.
- Some of your colors should be in a part-to-whole ratio of 1:3.

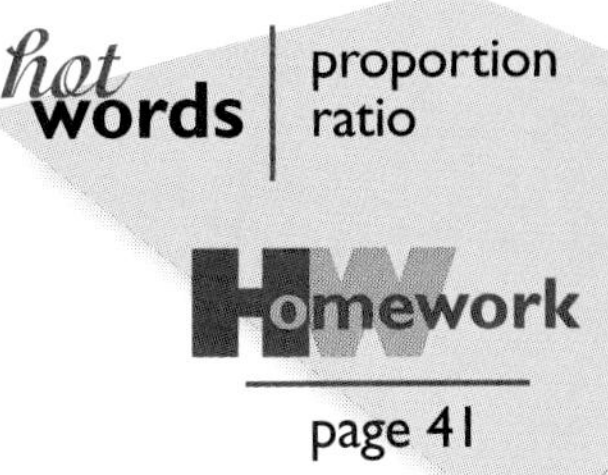

PHASE THREE

To: Staff Reporters
From: The Editors

Your final assignment is to plan an athletic banquet on a budget. A budget is the amount of money that is available for a particular use. When you complete the assignment you will write a magazine article for *Buyer Beware* on "Planning a Banquet on a Budget."

To prepare for this assignment, you will need to get some experience working with percents.

In this final phase you will use percents to help you become a smart shopper. Knowing about percents will help you to interpret discounts and resulting sale prices. If you know how much the discount actually reduces the price of an item, you can make informed decisions about what to buy and what not to buy.

Percents

WHAT'S THE MATH?

Investigations in this section focus on:

NUMBER

- Estimating expenses
- Exploring counting and rounding
- Calculating discounts
- Following a budget
- Determining savings
- Estimating and calculating percents

DATA and STATISTICS

- Interpreting a circle graph

GEOMETRY and MEASUREMENT

- Constructing a circle graph

Team Spirit

ESTIMATING PERCENTS

In this lesson you will use benchmarks and mental math to estimate the expenses of a football team. Then you will count and round to estimate the amount of money a field hockey team must raise to travel to a tournament.

Estimate Expenses

How can you use benchmarks to help you estimate expenses?

Estimating a percent of something is much the same as estimating a fractional part of something.

Last year the Cougars' football team's total expenses were $20,000. The table below shows what percent of the budget was spent on each expense. About how much money was spent on each item?

1 Make a table like the one shown below and complete the last column. Estimate the cost of each expense. Use benchmarks and mental math to make your estimates. Do not use paper and pencil.

2 The team spent exactly $800 on one expense. Which was it? Use estimation to figure it out.

3 The Tigers football team, cross-town rivals, spent 26% of its $16,000 budget on uniforms. Which team spent more on uniforms? Use estimation to figure it out.

Football Team Expenses

Items	Percent of Budget	Estimated Cost
Uniforms	23%	
Transportation	6%	
Coach's salary	48%	
Equipment	11%	
Officials' fees	4%	
Trainer's salary	8%	
	Total Expenses:	**$20,000**

Explore Counting and Rounding

How can counting and rounding help you estimate expenses?

A girls' field hockey team wants to play in a tournament. The school district will pay only a certain percentage of each expense. The team must raise the amount not paid by the district.

If you can find 50%, 10%, and 1% of a number mentally, you can also estimate some other percents mentally. For example, think of 25% as half of 50% and 5% as half of 10%.

Field Hockey Team Trip Expenses

Expense	Estimated Cost	Percent District Will Pay
Transportation	$700	3%
Meals	$600	12%
Hotel	$925	4%
Tournament fees	$346	22%
Tournament uniforms	$473	2.5%

1. Use counting and rounding to estimate the dollar amount the school district will pay for each expense.
2. Estimate the total amount the district will pay for all the expenses.
3. Estimate the total amount of money the team will need to raise.
4. Use your calculator to figure out the exact amount the team will need to raise for each expense.

Write About Your Estimates

Write about the estimates you made for the field hockey team expenses.

- What strategies did you use to estimate the expenses and total cost?
- What are some ways to determine whether your estimate is reasonable?

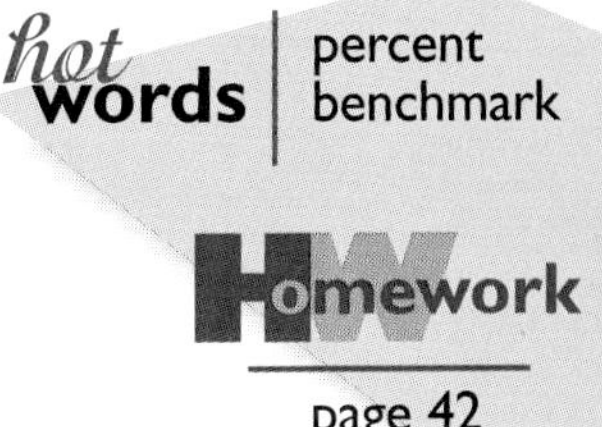

10 Playing Around

INTERPRETING AND CREATING CIRCLE GRAPHS

Circle graphs are useful tools for comparing percents. They show how different parts are related to a whole. In this lesson you will use a calculator to analyze data in a circle graph. Then you will use survey information about fund-raisers to make your own circle graph.

Use a Calculator to Find Percents in a Circle Graph

How can you analyze data in a circle graph?

The drama club plans to present a production of Moss Hart's *You Can't Take It With You.* The cost of the production is displayed in the following circle graph.

1 What is the total cost of the production? Use the circle graph to estimate what percent of the total cost was spent on each expense. Tip: Compare the size of each section to the whole circle.

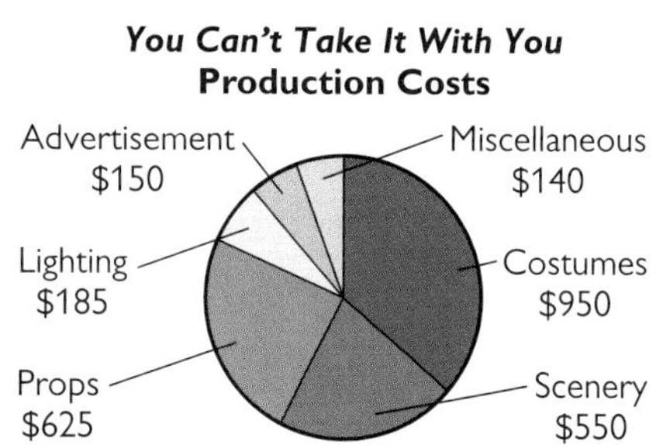

2 Use your calculator to help you find the percent of each expense in the circle graph. First, find the percent that will be spent on costumes by expressing the amount as a fraction; for example:

$$\frac{\text{cost of costumes}}{\text{total cost of production}}$$

Then, change the fraction into a decimal. Use your calculator to help you with this. Divide the cost of the costumes by the total cost of the production. To express the decimal as a percent, multiply the decimal by 100, round to the nearest whole number, and add a percent (%) sign.

3 Compare your estimates with the exact percent you got using the calculator. How close were your estimates?

Construct a Circle Graph

How can you construct a circle graph to show survey results?

The drama club decided it would like to attend a performance of Andrew Lloyd Webber's *Cats*. In order to see the musical production, they needed to raise the money to purchase the tickets. They conducted a survey in the middle school to find out which type of fund-raiser most students would be likely to attend. The survey gave the following results.

Fund-Raiser Choices by Number of Students in Each Grade

Choice	6th Grade Students	7th Grade Students	8th Grade Students	Total Students	Percent of Students
Carnival	88	75	69		
Raffle	45	49	41		
Car wash	66	54	52		
Bake sale	34	32	33		
Candy sale	12	11	15		
Don't know	3	4	1		

1. Figure out the total number of students who were surveyed.
2. Use your calculator to find the percent of students in the middle school that chose each activity. Round your answers to the nearest whole number.
3. Follow the guidelines on the handout Circle Graph to help you construct a circle graph to display your data.

Conduct Your Own Survey

Conduct a survey in your math class to find out which fund-raiser your classmates would choose to attend.

- Figure out what percent of the class chose each type of fund-raiser.
- Make a circle graph to represent your class's survey results.
- Compare your class's results to the data in the table.

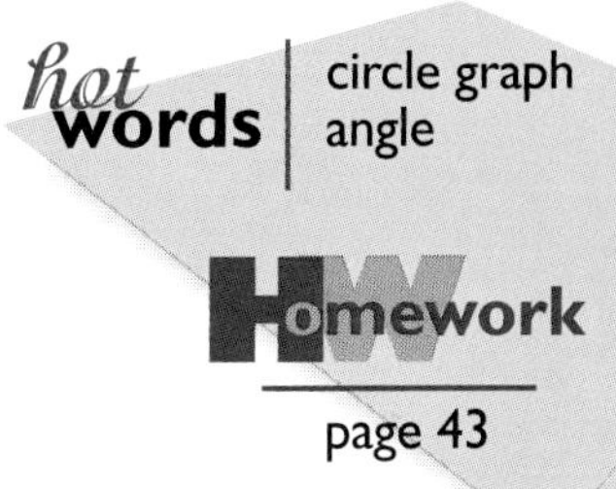

hot words
circle graph
angle
Homework
page 43

11 Sale Daze

USING PERCENTS TO CALCULATE SALE PRICES AND DISCOUNTS

Buying items on sale is a great way to save money. In this lesson you will calculate your savings by purchasing a skateboard at a percent-off sale. Then you will use discount coupons to shop for sporting equipment.

Determine Savings

How can you calculate your savings in a percent-off sale?

You want to join the after-school skateboard club this year, but you need to purchase the required equipment in order to participate. Skates on Seventh is advertising a percent-off sale. You have $175 to spend. Refer to this advertisement in answering the following questions.

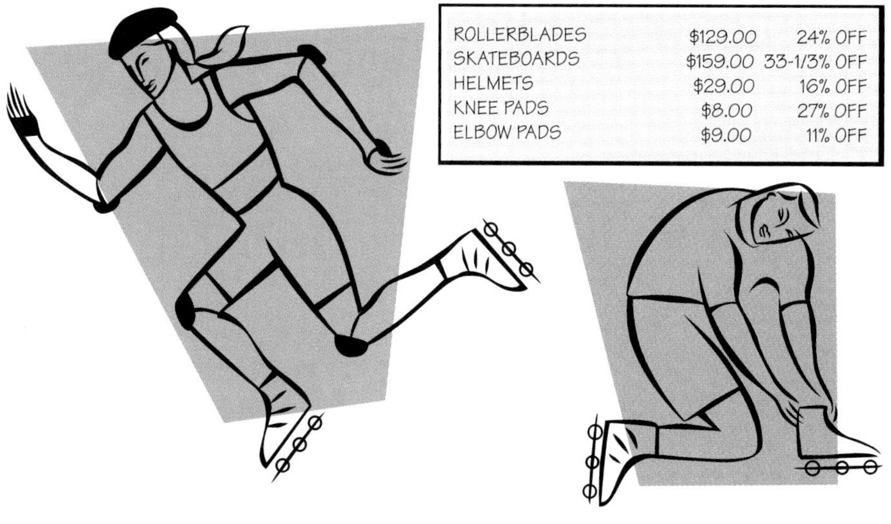

ROLLERBLADES	$129.00	24% OFF
SKATEBOARDS	$159.00	33-1/3% OFF
HELMETS	$29.00	16% OFF
KNEE PADS	$8.00	27% OFF
ELBOW PADS	$9.00	11% OFF

1. Make a quick estimate to see if you have enough money to buy a skateboard, helmet, knee pads, and elbow pads.
2. Use your calculator to figure out how much you will save for each item.
3. Figure out the sale price of each item.

Shop with Discount Coupons

How much money will you save by shopping with discount coupons?

The sports club wants to buy a variety of sports equipment for students to try out. They want you to buy as many new pieces of equipment as you can for their budget of $350. Fortunately, they have lots of discount coupons for you to use.

You need to buy *at least* three different kinds of equipment and use a different coupon from the handout Discount Coupons for each one. Remember, you can use each coupon only once and you can't use more than one coupon per type of equipment. The goal is to spend close to $350 without going over.

1. Decide which items you want to buy and which coupons you will use for each.
2. Make a table to show each original cost, the discount from the coupon, and the sale price.
3. Figure out the total cost of your purchases.
4. Figure out how much you saved by buying the items with discount coupons.

BASKETBALL	$23.00
CHAMPIONSHIP BASKETBALL	$52.00
VOLLEYBALL	$44.00
VOLLEYBALL NET	$109.00
CATCHER'S MITT	$51.50
CATCHER'S MASK	$15.70
BASEBALLS	$39.50 PER DOZEN
BASEBALL BAT	$26.95
BATTING HELMET	$11.85
SOCCER BALL	$26.30
TENNIS RACKET	$49.55
TENNIS BALLS	$12.00 FOR 3 CANS
FOOTBALL	$18.00
HOCKEY STICK	$49.00
HOCKEY PUCK	14.00 FOR 2

Design a Sale Advertisement

Create a colorful flyer announcing a sale at your favorite department store. For each advertised item, include:

- the original price
- the percent off, discount, and sale price

12 Percent Smorgasbord

CALCULATING PERCENTS IN REAL-LIFE SITUATIONS

A budget is a useful tool for keeping track of your money. In this lesson you will be given a set amount of money to plan an athletic banquet. You will need to plan a menu, select entertainment, and purchase awards, decorations, and gifts for the coaches. Finally, you will use all of your data to display your budget in a circle graph.

How can you work with a budget to make planning decisions?

Use a Budget to Plan an Athletic Banquet

You have been given $2,500 to plan the athletic banquet. The money was donated for the banquet and that's all it can be used for, so you need to spend close to $2,500. You can't spend more than this.

Athletic Banquet Attendance

Sport	Students	Coaches
Field hockey	20	2
Football	50	3
Soccer	30	2
Tennis	16	2
Basketball	20	2
Volleyball	12	1
Cross-country	12	1
Swimming	8	1
Lacrosse	18	2

1. Use the handout Athletic Banquet Price List to help you plan your choices for food, entertainment, awards, decorations, and gifts for the coaches.
2. When you have figured out how much you will spend for each category, write your plan for what you will be doing for each of the following: food, entertainment, awards, decorations, and gifts for the coaches.
3. Make a chart to record your total expenses in each category.
4. Write down the total amount you will spend on the banquet. Tell how much money, if any, you will have left over from the $2,500 you were given to spend.

Display Your Budget Data in a Circle Graph

You need to present the information in your banquet plan to the athletic advisory committee. Make a circle graph for the presentation.

- Use the circle on the handout Circle Graph.
- Use percents to label each sector of your circle graph.
- Make sure your sectors are clearly labeled and that the graph has a title.

Write an Article for *Buyer Beware*

Write an article for *Buyer Beware* magazine called "A Banquet on a Budget."

- Your article should include your chart and circle graph.
- Include tips to readers on how to make choices that save money.

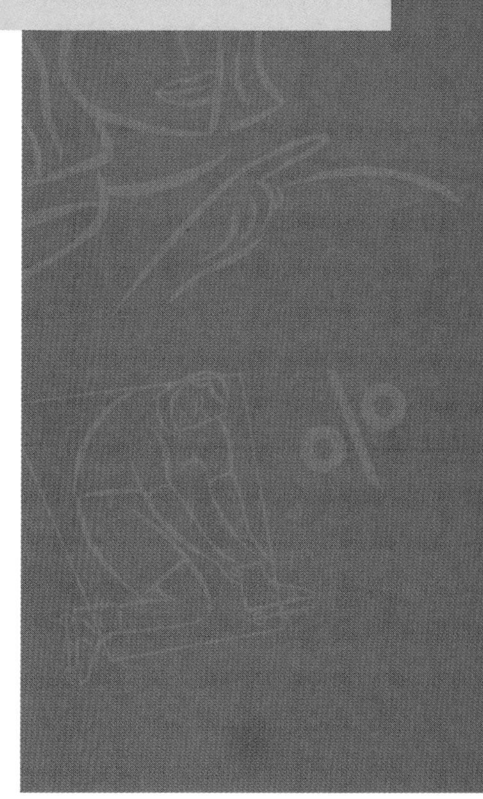

How can you make a budget presentation?

page 45

What's the Best Buy?

Applying Skills

Use your calculator to find the unit price for each of the following. Round to the nearest cent.

1. 7 oz of crackers for $1.19
2. 14 oz of cottage cheese for $1.19
3. 16 boxes of raisins for $5.60

Find the better buy based on unit price.

4. A 35-oz can of Best Brand Plum Tomatoes is on sale for $0.69. A 4-lb can of Sun Ripe Plum Tomatoes is $1.88.
5. A can of Favorite Dog Food holds 14 oz. Four cans are $1.00. The price of three cans of Delight Beef Dog Food, each containing 12 oz, is $0.58.
6. For each item, predict which is the better buy. Then use paper and pencil or a calculator to find the better buy.

	Item	Jefferson Auto Stores	Tom's Auto Parts
a.	oil	12 qt for $10.99	6 qt for $5.99
b.	antifreeze	12 oz for $3.79	6 oz for $1.79
c.	auto wax	6 cans for $14.29	5 cans for $12.98

Extending Concepts

7. Six cans of fruit drink are on sale for $1.95. Individually, the price of each can is $0.35. How much does Tanya save buying 6 cans on sale?
8. Tubes of oil paint can be bought in sets of 5 for $13.75 or bought separately for the unit price. What would be the price of 2 tubes of this oil paint?
9. The price of three bottles of Bright Shine Window Cleaner, each containing 15 oz, is $2.75. Two bottles of Sparkle Window Cleaner, each containing 18 oz, can be purchased for $1.98. Which is the better buy?

Writing

10. Create an advertisement for orange juice in which a small-size carton on sale is a better buy than a larger-size carton at regular price.

The Best Snack Bar Bargain

Applying Skills

The price graph below shows the unit prices for three different shampoos. Aloe Shampoo is $1.00 for 2 oz, Squeaky Clean is $2.75 for 3.5 oz, and Shine So Soft is $3.75 for 4 oz.

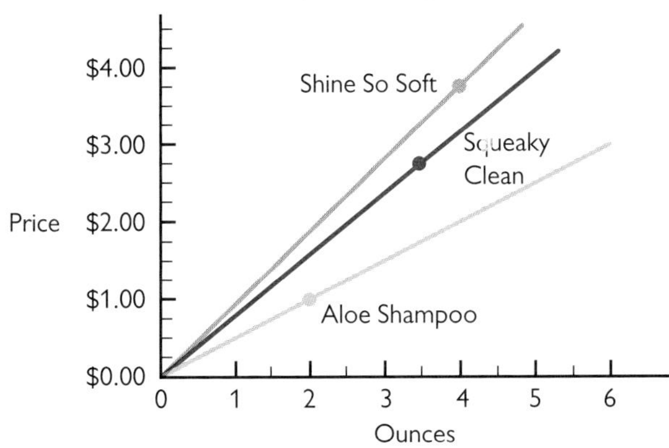

1. Use the graph to find the price of 4 oz of Aloe Shampoo.
2. Use the graph to find the price of 0.5 oz of Shine So Soft.
3. Use the graph to find out which shampoo has the lowest unit price.
4. Use the graph to find out which shampoo has the highest unit price.
5. Use your calculator to find the price of 1 oz of each shampoo. Use the graph to check your calculations.

Extending Concepts

Use the data for the products listed below to construct a price graph.

Product	Number of Units	Price
Crunchy Crackers	9.5 oz	$1.20
Buzzy Tuna	5.5 oz	$1.00
Bessy's Pancake Mix	12 oz	$1.60
Pino's Imported Pasta	8 oz	$2.00
Lean ground beef	9 oz	$2.40

Use the price graph you constructed to answer the following questions.

6. What is the price of 5 oz of Crunchy Crackers?
7. What is the price of 1 oz of Buzzy Tuna?
8. What is the price of 2 oz of Bessy's Pancake Mix?
9. What is the price of 14 oz of Pino's Imported Pasta?
10. What is the price of 15 oz of Lean Ground Beef?

Writing

The information below is missing data that is needed to complete a problem. Tell what might be missing. Make up data that could be used to complete and solve the problem.

11. The price of a box of biscuits is $0.89. On the box it says, "New larger size—15 ounces."

Cheaper By the Pound

Applying Skills

Find the price per pound to decide the better buy.

Potatoes	
2 lbs	$1.09
5 lbs	$2.69
10 lbs	$3.59
20 lbs	$6.29

1. What is the price per pound of each package of potatoes?

2. Which size package of potatoes is the best buy in terms of price per pound?

Calculate the price per pound of the items below.

	Item	Price per Item	Weight in Pounds	Price per Pound
3.	Bike	$179.99	45	
4.	Rollerblades	$135.99	11	
5.	Basketball	$ 24.99	2	
6.	1996 complete set of baseball cards	$ 26.99	3.6	
7.	Earrings	$ 16.00	0.25	
8.	Watch	$ 34.95	0.25	
9.	Pearl ring	$ 79.99	0.13	

10. For which of the items above would you pay the least per pound?

Extending Concepts

11. Alejandro bought an 18-lb watermelon for $4.00. To the nearest cent, what is the price per pound?

12. A 5-lb bag of dog food sells for $3.85. Maurice's dog eats 2 bags of dog food every month. What is the monthly price per pound of the dog food?

Making Connections

13. Select several magazines or newspapers. Find out how much a subscription costs to each of the magazines or newspapers. Compare the unit price to the newsstand price.

It Really Adds Up

Applying Skills

Crunchy Popcorn	No-Ad Popcorn
$0.45 per bag	$0.35 per bag

Find the price of buying one bag of popcorn every day for a week.

1. Crunchy Popcorn

2. No-Ad Popcorn

Find the price of buying one bag of popcorn every day for a month (4.3 weeks).

3. Crunchy Popcorn

4. No-Ad Popcorn

Find the price of buying one bag of popcorn every day for a year.

5. Crunchy Popcorn

6. No-Ad Popcorn

7. How much would you save if you bought No-Ad Popcorn instead of Crunchy Popcorn for:

a. 1 week

b. 1 month (4.3 weeks)

c. 1 year

8. If you bought a bag of No-Ad Popcorn every day instead of Crunchy Popcorn, how many days would it take to save $30? $75? Explain how you figured it out.

Extending Concepts

9. At Jacy's Market, you can get 5 mangos for $1.95. At Nia's Market, you can get 3 for $1.29. Are the mangos cheaper at Jacy's or at Nia's?

10. You can get 3 cans of Mei Mei's Soup for $1.23 and 2 cans of Pei's Soup for $0.84. Which brand costs less per can?

11. Why would a store owner price an item at $9.99 for 5 instead of $2.00 each?

Making Connections

12. Select two supermarket advertisements from a newspaper. Compare the prices of similar items. Which store seems to have the better buys? Give reasons for your answer.

Quilting Ratios

Applying Skills

Write each ratio as a fraction in lowest terms.

1. 6 to 8 **2.** 8:44

3. $\frac{60}{32}$ **4.** 20 to 30

Write two ratios that are equal to each ratio.

5. 5:30 **6.** 12:15

7. 30:12 **8.** 8:2

Using the quilt below, write each ratio as a fraction in lowest terms. Then tell whether it is a part-to-part or part-to-whole ratio.

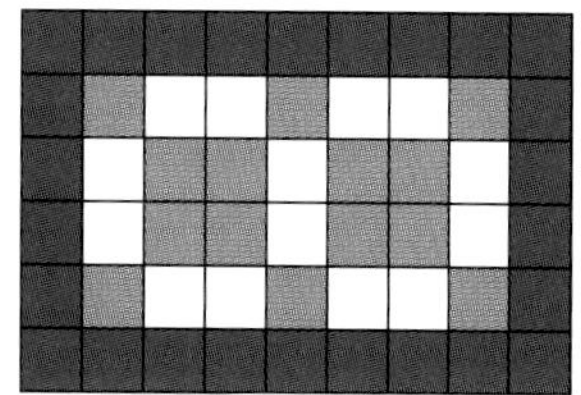

9. light blue squares to dark blue squares

10. dark blue squares to all squares

11. light blue squares to all squares

Extending Concepts

The data below show how some students spent their time from 4 P.M. to 5 P.M. yesterday. Decide if statements **12–15** are true or false.

How Students Spent Their Time

	Number of Students
Homework	𝍸 𝍸
Sports practice	𝍸
Music practice	𝍸
Chores or job	𝍸 I
Other	IIII

12. One out of every three students did homework.

13. One out of every five students did chores.

14. The ratio of students doing homework to students practicing music is 5 to 2.

15. The ratio of students doing chores to students practicing music or sports is 2 to 3.

Making Connections

16. Sports statistics use ratios to describe a player's performance. Choose several players in a sport you enjoy. Research the players' statistics and write several ratios for each set of statistics. For example, baseball ratios might include hits to at bats or total bases to hits. Football ratios could include field goals made to field goals attempted.

Compare the ratios for players you have chosen. Do the ratios explain why one player is more valuable than another?

In the Mix

Applying Skills

Use any method to solve each problem.

1. This pattern shows one "octave" on a piano keyboard.

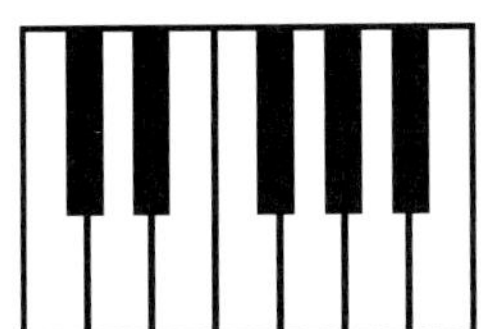

a. Electronic keyboards come in different sizes. One keyboard has 4 octaves. How many black keys does it have?

b. Amy's favorite keyboard has 35 white keys. How many black keys does it have?

2. The scale on a map is 3 inches = 4 miles.

a. How far is the actual distance between two towns that are 10.5 inches apart on the map?

b. How many inches apart are two streets that actually are one mile apart?

Use ratio tables to solve each problem.

3. Anya's class is selling wrapping paper. For every 5 rolls they sell, they make a profit of $1.80. She wants to figure out how many rolls the class needs to sell to make a profit of $270. Fill in the missing numbers in her ratio table.

Number of rolls	5	10	15	30	**d.**
Profit	$1.80	**a.**	**b.**	**c.**	$270

4. A passenger jet travels at an average speed of 450 miles per hour. Use the ratio table to find the missing times or distances.

Time (hours)	1	2	4	**c.**	10
Distance (miles)	**a.**	**b.**	1,800	2,700	**d.**

Extending Concepts

5. The price of first-class postage is 37¢ for up to one ounce, 54¢ for up to two ounces, 71¢ for up to three ounces, 88¢ for up to four ounces, and $1.05 for up to five ounces.

Marco claims the table below is a ratio table. Is he correct? Explain.

Weight (ounces)	1	2	3	4	5
Cost	37¢	54¢	71¢	88¢	$1.05

Making Connections

6. In question 5, the number of ounces increases by one, and the cost increases 17¢. Even though there is an adding pattern in both sets of numbers, the numbers do not stay in the same ratio to each other.

Describe another real-life example that has a constant adding pattern but does not stay in the same ratio to each other.

Halftime Refreshments

Applying Skills

Determine whether each pair of ratios is proportional. Write *yes* or *no*.

1. $\frac{3}{4}, \frac{9}{16}$
2. $\frac{4}{6}, \frac{6}{9}$
3. $\frac{20}{16}, \frac{15}{12}$
4. $\frac{7}{12}, \frac{8}{15}$

Solve each proportion.

5. $\frac{4}{6} = \frac{n}{21}$
6. $\frac{12}{n} = \frac{8}{15}$
7. $\frac{n}{28} = \frac{30}{14}$
8. $\frac{4}{n} = \frac{n}{16}$

Write a proportion and solve each problem.

9. Merchants price their products based on proportions. Suppose 12 cans of soda cost \$4.80. What is the price of 36 cans of the same soda?

10. A tree casts a shadow that is 28 feet long. Liza notices that at the same time, her shadow is 3 feet long. She knows that she is $5\frac{1}{2}$ feet tall. If the ratio of the height of the tree to the length of its shadow is proportional to the ratio of Liza's height to the length of her shadow, how tall is the tree?

11. There are onions and green peppers in a bag of frozen mixed vegetables. The ratio of ounces of onions to ounces of green peppers is 4 to 9. How many ounces of onions are there if there are 30 ounces of green peppers?

12. Suppose you buy 2 CDs for \$21.99. How many CDs can you buy for \$65.97? Assume all of the CDs cost the same amount.

Extending Concepts

13. How many proportions can you make using only the numbers 1, 3, 4 and 12?

14. At Friday's football game, the athletic boosters sold three times as many hot dogs as brownies. Altogether, 500 hot dogs and brownies were sold. How many of each item were sold?

15. **Challenge** Which plot of land is most square? Explain your answer.

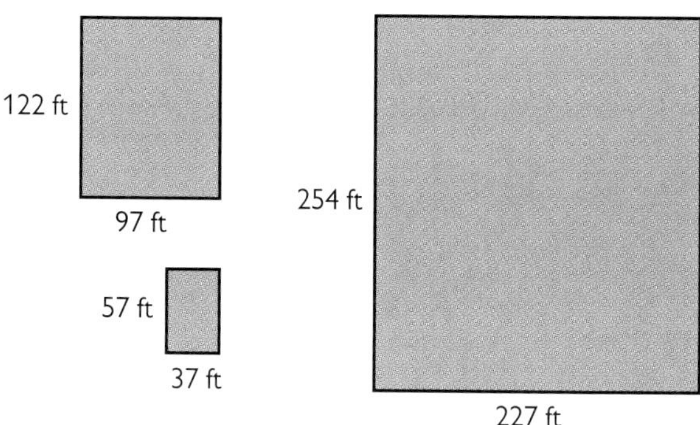

Making Connections

16. Explore exchange rates in currency (money) from different countries.

    a. Choose three different countries. Find out the current exchange rate between the currency used in that country and American dollars.

    b. For each country, set up a proportion to calculate how much money from that country you would get for \$100 American dollars.

Can I Use a Proportion?

Applying Skills

Determine whether the following sets of numbers are proportional to each other. Write *yes* or *no*. Explain your reasoning.

1.

1st number	2	4	6	8	10
2nd number	4	6	8	10	12

2.

1st number	3	6	9	12	15
2nd number	4	8	12	16	20

3.

1st number	2	3	5	8	12
2nd number	1	1.5	2.5	4	6

4.

1st number	0	3	4	8	12
2nd number	2	5	6	10	14

5. For each of the problems above, make a line graph of the table of values. You should end up with four separate lines.

6. Two of the problems above have numbers that are proportional. What is true about the line graphs of these two problems?

7. Two of the problems above have numbers that are not proportional. How are the line graphs of these two problems different from the line graphs of the proportional sets of numbers?

Extending Concepts

8. Jack is 7 years younger than his sister Nancy. This year on their birthdays, he turned 7 and she turned 14. Nancy noticed that she was twice as old as Jack. She wondered, "Will I ever be twice as old as Jack again? If so, when?"

a. Make a table of their ages for the next five years.

b. For each year, calculate the ratio of Nancy's age to Jack's age as a single number rounded to the nearest hundredth. What do you notice about the ratios?

c. When will Nancy be twice as old as Jack again? Explain your answer.

d. Is there a time when Nancy will be one and a half times as old as Jack? If so, when? If not, explain why not.

e. Will the ratio of their ages ever become 1? Explain why or why not.

Writing

9. Robert wants to make hot cocoa for his friends. He knows he needs 2 spoonfuls of cocoa mix for each mug of hot cocoa that he makes. Robert thinks: "I'll make sure the number of spoonfuls is always one more than the number of mugs I want to make." Do you agree with Robert's reasoning? Explain why or why not.

Team Spirit

Applying Skills

The field hockey team's total expenses for last year were $10,000. Estimate how much was spent on each of the expenses listed below. Use benchmarks and mental math to make your estimates.

	Item	Percent of Budget	Estimated Cost
1.	Uniforms	22%	
2.	Transportation	4%	
3.	Coach's Salary	12%	
4.	Equipment	49%	
5.	Officials' Fees	6%	
6.	Trainer's Salary	7%	

Estimate each number in items **7–10.** Then use your calculator to see how close your estimate is.

7. 49% of 179

8. 24% of 319

9. 19% of 354

10. 34% of 175

11. Find 7% of $400.

7% means________ for every________

12. Find 12% of $300

12% means________ for every________

Extending Concepts

13. To estimate 24% of 43, LeRon substituted numbers and found 25% of 44. His answer was 11. Using his calculator, he found that the exact answer is 10.32. LeRon concluded that substituting numbers causes you to overestimate. Do you agree? If not, give a counterexample.

14. Nirupa calls home from college at least once a week. A 30-minute phone call costs $10.00 on weekdays. Nirupa can save 20% if she calls on a weekend. How much money does she save on a 30-minute call made on Saturday?

Making Connections

15. Look through newspapers and magazines to find articles involving percents. Design a collage with the articles. Write out percents from 1 through 100 and their equivalent fractional benchmarks.

Playing Around

Applying Skills

The circle graph below shows the budget for the middle school production of *The Music Man.*

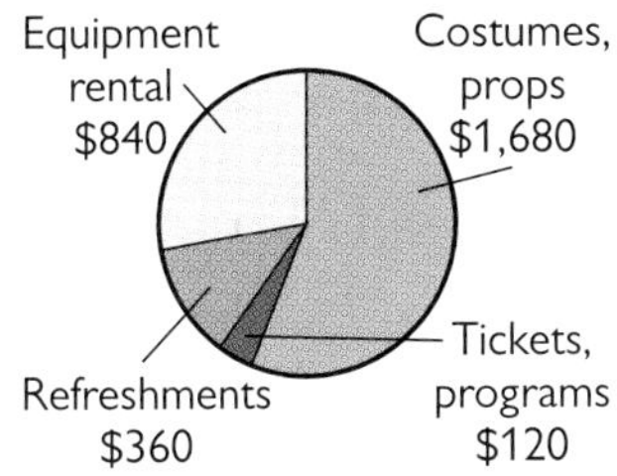

Use the circle graph to estimate what percent of the total cost was spent on each of the following:

1. costumes and props

2. tickets and programs

3. refreshments

4. equipment rental

Use your calculator to help you find the actual percent of the total cost that was spent on each of the following:

5. costumes and props

6. tickets and programs

7. refreshments

8. equipment rental

Use the following information to make a circle graph.

9. Ticket sales for *The Music Man* totaled $560. Students collected the following amounts: Vanessa $168, Kimiko $140, Ying $112, Felicia $84, and Norma $56. Label the circle graph, using names and the percents collected. Give the graph a title.

Extending Concepts

Step-in-Time shoe store took in the following amounts in January:

Men's dress shoes	$750
Women's dress shoes	$1,500
Children's sneakers	$2,000
Adult athletic shoes	$850

The circle graph below was made using the information above.

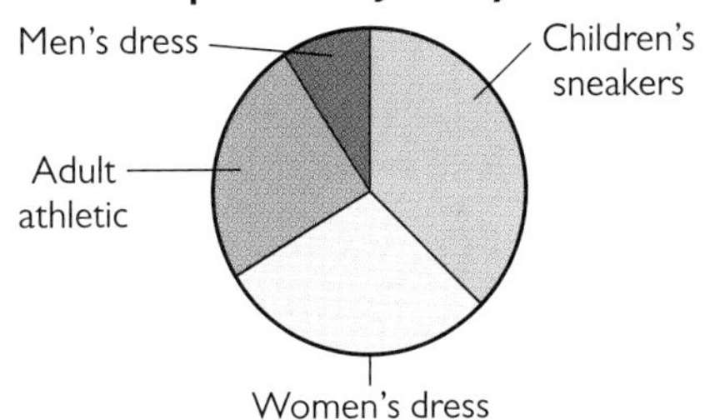

10. Was the circle graph made correctly? Explain.

Making Connections

Percentage of the World's Population by Continent

Continent	Percentage
Africa	12%
Europe	14%
North America	8%
Asia	60%
South America	6%

Source: *World Almanac 1996*

11. Use the handout Circle Graph to make a circle graph of the world's population.

Sale Daze

Applying Skills

Use a calculator to find the discount and sale price for items 1–4. Round to the nearest cent.

1. regular price: $78
rate of discount: 25%

2. regular price: $29
rate of discount: 20%

3. regular price: $45
rate of discount: 15%

4. regular price: $120
rate of discount: 37%

Which shop has the better buy?

	Item	Super Sports	Sammy's Sport Shop
5.	Football	Regular price: $39.95 rate of discount: 10%	Regular price: $42.95 rate of discount: 15%
6.	Basketball	Regular price: $36.50 rate of discount: 20%	Regular price: $35 rate of discount: 15%
7.	Helmet	Regular price: $19.95 rate of discount: 20%	Regular price: $17.95 rate of discount: 15%

8. Orit needed a helmet to skate on the half-pipe at Skateboard Plaza. She bought one for 60% off the regular price of $31.50. How much did she save? How much did she pay?

9. Salvador paid $27.50 for a catcher's mask that was on sale. The regular price was $36.90. What was the discount?

Extending Concepts

10. Kai-Ju has saved $40 to buy a tennis racket that regularly sells for $59.99. She read an ad that announced a 25% discount on the racket. Has she saved enough money to buy it? If so, how much will she have left over? If not, how much more does she need?

11. Ryan went into a hardware store that had a 20%-off sale. His two purchases originally had prices of $38.75 and $7.90. How much did Ryan save because of the sale?

Writing

Make up the missing data in items 12 and 13. Then write a word problem that can be solved using the data.

12. roller skates
regular price: $89.99
rate of discount:

13. tennis racket
regular price: $45.89
rate of discount:

Percent Smorgasbord

Applying Skills

For items 1–4, how many items on the list can the shopper buy without overspending?

1. shopper has $140
discount: 12%

video game set	$89.95
video game cartridge	$29.50
blow dryer	$27.50
sweater	$34.00

2. shopper has $150
discount: 10%

roller skates	$99.00
compact disk	$12.99
jacket	$75.00
shirt	$18.00

3. shopper has $180
discount: 25%

bicycle	$129.00
bike helmet	$ 39.00
tennis racket	$ 89.00
ring	$ 58.00

4. shopper has $62
discount: 50%

radio	$29.00
jeans	$32.00
earrings	$18.00
shoes	$44.00

5. A box of Munchy Cereal contains 24 oz of cereal. It is on sale for 50% off the regular price of $4.80. Toasty Cereal, which contains 50% more cereal than Munchy Cereal, is $4.80 per box. Which box has the lower price per ounce? To the nearest cent, how much less is the price per ounce for this box?

Extending Concepts

Use your calculator to find the total for each bill. The tip is 15%.

6. **South of the Border**

Beef tacos	$6.25
Guacamole	$4.50
Chicken tamale	$7.50

7. **Thai Cuisine**

Chicken with ginger	$7.95
Sweet and sour chicken	$8.90
Shrimp and baby corn	$9.95

Writing

8. Answer the letter to Dr. Math.

Dear Dr. Math,
Neely's Hot Dogs advertises that their hot dogs contain more protein than fat. Their hot dogs contain proteins, carbohydrates, and fat. Each hot dog contains 11 g of fat and 2 g of carbohydrates. This makes up 55% of the hot dog's content. Could the advertisement be true? How can I tell?
Lois Kallory

The McGraw-Hill Companies

This unit of MathScape: Seeing and Thinking Mathematically was developed by the Seeing and Thinking Mathematically project (STM), based at Education Development Center, Inc. (EDC), a non-profit educational research and development organization in Newton, MA. The STM project was supported, in part, by the National Science Foundation Grant No. 9054677. Opinions expressed are those of the authors and not necessarily those of the Foundation.

CREDITS: Unless otherwise indicated below, all photography by Chris Conroy and Donald B. Johnson.

3 (tl)SuperStock, (tc)Photodisc/Getty Images; **14 21** Photodisc/Getty Images.

Send all inquiries to:
Glencoe/McGraw-Hill
8787 Orion Place
Columbus, OH 43240-4027

ISBN: 0-07-866806-9

2 3 4 5 6 7 8 9 10 058 06 05 04